baylands

valley floor

VISION
for a resilient Silicon Valley landscape

San Francisco Estuary Institute
April Robinson
Erin Beller
Robin Grossinger
Letitia Grenier

June 2015

SAN FRANCISCO ESTUARY INSTITUTE
AQUATIC SCIENCE CENTER

SFEI
A○S○C

SFEI Publication # 753

This document presents a preliminary vision for landscape resilience across the streams, hills, baylands, and urban areas of Silicon Valley. It is a product of Resilient Silicon Valley, a project of the San Francisco Estuary Institute to create a science-based vision for ecosystem health and resilience in Silicon Valley (resilientsv.sfei.org).

The vision outlined here was developed by applying a set of resilience principles (Beller et al. 2015) to Silicon Valley, in collaboration with a team of regional science advisors, to identify landscape elements that are likely to contribute to resilience in the region. It is intended to provide a broad foundation for restoration and management strategies and contribute to discussions amongst scientists, planners, managers, and other stakeholders about specific actions that would improve landscape resilience. However, this document is not intended to provide on-the-ground recommendations, and the vision elements presented here will need to be made more spatially specific, quantitative, and aligned with current planning efforts before they can be implemented in a meaningful way.

Introduction

Silicon Valley: more than just a place, the phrase is synonymous across the globe with high-tech innovation and creativity. Silicon Valley is a thriving urban area, an incubator for cutting-edge ideas and home to millions of people. At the same time, alongside the tech campuses and subdivisions, there are dozens of rare and endemic plants and animals that call Silicon Valley home. From the hills to the bay, Silicon Valley supports a diverse array of habitats: redwoods blanket mountainsides, egrets nest in trees amidst busy commercial areas, and steelhead make their way up streams bisecting suburban cities. The environmental quality of our region is a recognized component of its attractiveness and success, yet many elements of the local ecosystem have either been lost or are fragmented and vulnerable.

Over the coming decades, the Silicon Valley landscape will inevitably change and evolve in significant ways. Buildings will be renovated and redeveloped, flood-control channels will be redesigned, and new parks will be created, among innumerable other changes to infrastructure and landscapes. Each of these modifications offers an opportunity to help realize the region's enormous ecological potential by contributing to the creation of biodiverse, healthy ecosystems across the Silicon Valley. But which actions should receive priority? How can we ensure that our actions add up to something ecologically meaningful and lasting? How do we create landscapes that are resilient – that is, that have the capacity to persist and evolve over time, even as conditions change?

Current and anticipated stressors to Silicon Valley ecosystems make these questions even more challenging and pressing. The population of Silicon Valley is expected to grow substantially in the coming decades, placing increasing demands on the region's natural resources and increasing pressure on remaining open space. Climate change and associated stressors – including sea level rise, increased drought intensity, increased air temperatures, and increased storm intensity and associated flood risk – add additional complexity to ecosystem management. In the face of this uncertain future, it is imperative that we promote landscapes that support the species, habitats, and ecosystems likely to successfully adapt, thrive, and be as self-sustaining as possible over time.

The unique ecological resources of Silicon Valley, coupled with its profound creative and financial capital, have the potential to make the region a hub for ecological as well as technological innovation. Yet moving forward in a way that supports biodiverse, functioning ecosystems alongside population and economic growth will require thoughtful and courageous planning and action. We will need a shared, long-range vision for a resilient Silicon Valley in order to take full advantage of opportunities to shape our landscapes. This document takes one step toward that vision by outlining a preliminary description of the landscape elements needed to bolster ecological resilience at a landscape scale (hereafter "landscape resilience") in Silicon Valley.

About this document

PROJECT GOALS AND SCOPE

The goal of this document is to envision the key landscape elements that are likely to contribute to landscape resilience. By "landscape resilience," we mean the ability of a landscape to sustain native biodiversity, ecological functions, and critical physical processes over time, in the face of climate change, urbanization, and other stressors. It is important to note that our goal is not necessarily the resilience of Silicon Valley ecosystems as they currently are – that is, with fragmented, often-degraded habitats and often-low capacity for persistence and adaptation. Rather, we aim to envision a resilient Silicon Valley landscape that would support high levels of desired ecological functions and biodiversity over time, even as some transformations in landscape structure and condition occur. While many of Silicon Valley's species, habitats, and ecological functions have suffered from past and current land use practices, we believe that planning for future land use changes can help ensure the resilience of these ecosystems in the future.

This document is a product of Resilient Silicon Valley, a project of the San Francisco Estuary Institute to create a science-based vision for ecosystem health and resilience in Silicon Valley (resilientsv. sfei.org). Resilient Silicon Valley's geographic scope includes the Santa Clara County watersheds that drain to the San Francisco Bay, from the hills down to the baylands (including urbanized areas, but excluding southern Santa Clara Valley; fig. 1). In particular, we focus on the ecosystems of the

KEY DEFINITIONS

Landscape resilience is the ability of a landscape to sustain native biodiversity, ecological functions, and critical physical processes over time, in the face of climate change, urbanization, and other stressors.

Biodiversity includes the variety of life at all levels, from genes to ecosystems; it is supported by ecological functions and physical processes.

Ecological function refers to all the ways that ecosystems support life (e.g., supporting complex native food webs, providing food resources, functioning as movement corridors, providing shade, providing nesting sites, and attenuating wave action).

region's four major landscape units: creeks, the hills, the urbanized valley floor, and baylands. We focus here on describing the elements of a resilient Silicon Valley, rather than prescribing actions for how to achieve that vision. For more information on recommended actions see "How do we get there," page 32).

DEVELOPING THE VISION

In an earlier stage of Resilient Silicon Valley we developed a Landscape Resilience Framework (Beller et al., 2015), which served as a guide for this vision. That framework identifies seven fundamental principles of landscape resilience (see call-out box, page 5), along with the most relevant elements within each principle that relate to planning, restoration, conservation, and management actions. To produce the list of elements for a resilient Silicon Valley, we created a series of worksheets that guided us through an initial application of these principles to Silicon Valley (see Appendix, page 36). We then collaborated with a team of regional science advisors to synthesize the most important elements from the worksheets into the lists presented here (Table 1).

These elements are grounded in an understanding of Silicon Valley's unique ecological characteristics: its geophysical context; past, present, and potential ecosystems; and human history and land use trajectories. These characteristics help identify locally appropriate landscape characteristics by providing insight into the potential priorities and opportunities as well as the constraints of this specific place. We endeavored to think big-picture and at long

Table 1. Regional Science Advisory Team members and affiliations.

Advisor	Affiliation
David Ackerly	UC Berkeley
Peter Baye	Independent Consultant
John Bourgeois	State Coastal Conservancy
Josh Collins	San Francisco Estuary Institute
Andy Collison	ESA PWA
Ron Duke	H.T. Harvey & Associates
Nicole Heller	Pepperwood Foundation
Rob Leidy	U.S. Environmental Protection Agency
Jeremy Lowe	San Francisco Estuary Institute
Lisa Micheli	Pepperwood Foundation
Bruce Orr	Stillwater Sciences
Steve Rottenborn	H.T. Harvey & Associates
Dan Stephens	H.T. Harvey & Associates

time scales within the context of Silicon Valley's current land uses: that is, we assumed that Silicon Valley will stay urbanized, and also that there will be many opportunities for landscape redesign as major infrastructure is replaced in the coming decades.

This document represents a first exploration of how the Landscape Resilience Framework might be applied to construct a vision for a resilient Silicon Valley landscape. We anticipate that it will serve as a catalyst for subsequent discussions among scientists, planners, managers, and other stakeholders about specific landscape elements that would improve resilience and could be integrated into future conservation and management documents. At this stage, we do not yet identify a plan for actions, locations, or design elements that suggest how the vision could be implemented, nor do we specify numerical targets or landscape metrics that would serve as a basis for measuring progress and success. In subsequent phases of Resilient Silicon Valley, we will translate this vision into more spatially explicit and metrics-based recommendations and plans. This will be coupled with more extensive scientific and stakeholder involvement that will allow us to make the vision more robust, collaborative, and quantitative; ultimately, we hope it will contribute guidance to local planning documents such as Master Plans and General Plans.

WHAT ABOUT BENEFITS TO PEOPLE?

In addition to the benefits that managing for landscape resilience would confer to wildlife, many of the elements outlined in this vision will undoubtedly yield important co-benefits to people and society: for example, providing services such as flood protection, clean water, groundwater filtration and recharge, and enhancing human access to nature. A Silicon Valley with increased landscape resilience is also likely to provision many of these ecosystem services in a more self-sustaining and cost-effective manner than a landscape with lower resilience. Our goal in emphasizing benefits to non-human species is to develop a vision of the resilience of ecosystems as one crucial component of a broader vision for social-ecological resilience in Silicon Valley.

Of course, this cannot be achieved in a vacuum, and there are many additional dimensions of resilience — including the resilience of economic and social institutions, of infrastructure, and of ecosystem services — that must be taken into account in order to effectively implement a vision for landscape resilience, along with considerations such as cost and competing land uses. Other planning processes, most notably Silicon Valley 2.0, provide a more detailed, overarching vision for Silicon Valley's interrelated social, economic, and natural systems.

PRINCIPLES OF LANDSCAPE RESILIENCE

These principles were developed as part of the Landscape Resilience Framework and inform the Resilient Silicon Valley Vision presented in this document.

1 — SETTING ·········▸ Unique geophysical, biological, and cultural aspects of a landscape that determine potential constraints and opportunities for resilience

2 — PROCESS ·········▸ Physical, biological, and chemical drivers, events, and processes that create and sustain landscapes over time

3 — CONNECTIVITY ·········▸ Linkages between habitats, processes, and populations that enable movement of materials and organisms

4 — DIVERSITY & COMPLEXITY ·········▸ Richness in the variety, distribution, and spatial configuration of landscape features that provide a range of options for species

5 — REDUNDANCY ·········▸ Multiple similar or overlapping elements or functions within a landscape that promote diversity and provide insurance against loss

6 — SCALE ·········▸ The spatial extent and time frame at which landscapes operate that allows species, processes, and functions to persist

7 — PEOPLE ·········▸ The individuals, communities, and institutions that shape and steward landscapes

Relationship to other planning efforts

This vision was informed by a number of existing planning documents that identify regional recommendations for land and water management in Santa Clara Valley. These include the *Baylands Ecosystem Habitat Goals Update* for bayland habitats, the *Conservation Lands Network* for upland habitats, and the *Santa Clara Valley Greenprint* and *Santa Clara Valley Habitat Plan* for overall conservation and management across the valley (Table 2). These documents articulate visions for many aspects of the Santa Clara Valley landscape; the vision presented here incorporates many recommendations from across these various efforts.

The scope of the Resilient Silicon Valley project differs from that of existing efforts in three main ways. First, this project differs in its explicit focus on the *resilience and adaptation* of native ecosystems as a whole (i.e., rather than on specific species or communities). Second, it is distinguished by its focus on *ecological function and biodiversity,* rather than on implementation, social and economic considerations, infrastructure, recreation, or ecosystem services. Lastly, it aims to consider these goals at a *landscape scale,* integrating across terrestrial, aquatic, tidal, and urban areas. While existing documents cover aspects of these considerations, none address them all. For example, Silicon Valley 2.0 focuses on climate adaptation with a broader topical purview and a less detailed focus on ecological function and biodiversity. Similarly, the *Baylands Ecosystem Habitat Goals Update* focuses on ecological function, but only for a portion of the Santa Clara Valley landscape.

The Resilient Silicon Valley project provides greater resolution on the resilience of ecological functions, supplying details that could nest within broader efforts such as Silicon Valley 2.0. In addition, its explicit focus on resilience provides support and rationale for many of the recommendations in other regional planning documents while also identifying potential gaps, adding detail, and forging links across them.

Table 2. Key relevant regional planning documents for Silicon Valley (see References for complete citations).

Planning Document	Date Published	Lead Agency	Spatial coverage	Topical focus	Website
Vision for a Resilient Silicon Valley Landscape	**2015**	**San Francisco Estuary Institute**	**Santa Clara Valley (excluding South County)**	**ecological resilience at a landscape scale, integrated across hills, urban areas, creeks and wetlands, and baylands**	**resilientsv.sfei.org/**
Silicon Valley 2.0	forthcoming	Santa Clara County	Santa Clara Valley	climate change adaptation strategies for economic, social, and environmental assets	www.sccgov.org/sites/osp/SV2/Pages/SV2.aspx
Integrated Water Resources Master Plan	forthcoming	Santa Clara Valley Water District	Santa Clara Valley	Water supply, flood protection, water quality, ecological resources	www.valleywater.org/iwrmp/
Baylands Ecosystem Habitat Goals Update	forthcoming (original 1999)	State Coastal Conservancy	SF Bay (baylands)	baylands ecological function	—
Santa Clara Valley Greenprint	2014	Open Space Authority	Santa Clara Valley (excluding West County)	wildlands, water resources, agriculture and ranchlands, recreation and education	www.openspaceauthority.org/about/strategicplan.html
Deep Roots, Green Future	2014	Committee for Green Foothills	Santa Clara Valley	natural communities for people and wildlife	www.greenfoothills.org/vision/
Santa Clara Valley Habitat Plan	2012	Santa Clara Valley Habitat Agency	Santa Clara Valley (excluding West County)	species, community, and landscape-level conservation	scv-habitatagency.org/
Conservation Lands Network	2011	Bay Area Open Space Council	SF Bay Area (focus on uplands and creeks)	protected lands and open space, biodiversity and habitat, water resources, people and conservation	www.bayarealands.org/

Environmental setting

Santa Clara Valley, also known as Silicon Valley, is located in Santa Clara County in the central California Coast Range, nestled between the Santa Cruz Mountains to the west, the Diablo Range to the east, and San Francisco Bay to the north. The valley was named *Llano de los Robles,* or Plain of the Valley Oaks, by the earliest Spanish explorers in the 1700s, so called for the emblematic oak groves that stretched for miles on alluvial valley soils. The valley supported a diverse mosaic of habitats in addition to oak savannas and woodlands, from grassland, chaparral, and forests in the hills and on the upper valley floor to the extensive tidal marshlands ringing the bay (Grossinger et al. 2006, 2007; Beller et al. 2011; fig. 2). Despite the relatively dry Mediterranean climate, high groundwater in many places supported expansive wetland habitats in flat, low-lying areas, including seasonal and perennial meadows, ponds, and marshes; willow groves; and riparian forests.

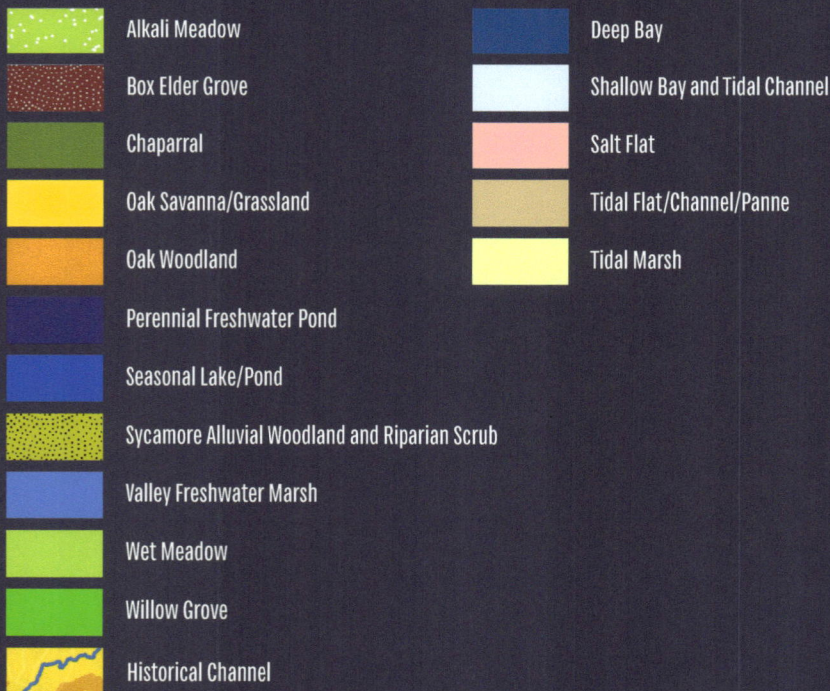

SAN FRANCISCO BAY

Alkali Meadow

Box Elder Grove

Chaparral

Oak Savanna/Grassland

Oak Woodland

Perennial Freshwater Pond

Seasonal Lake/Pond

Sycamore Alluvial Woodland and Riparian Scrub

Valley Freshwater Marsh

Wet Meadow

Willow Grove

Historical Channel

Deep Bay

Shallow Bay and Tidal Channel

Salt Flat

Tidal Flat/Channel/Panne

Tidal Marsh

N

1:250,000

5 miles

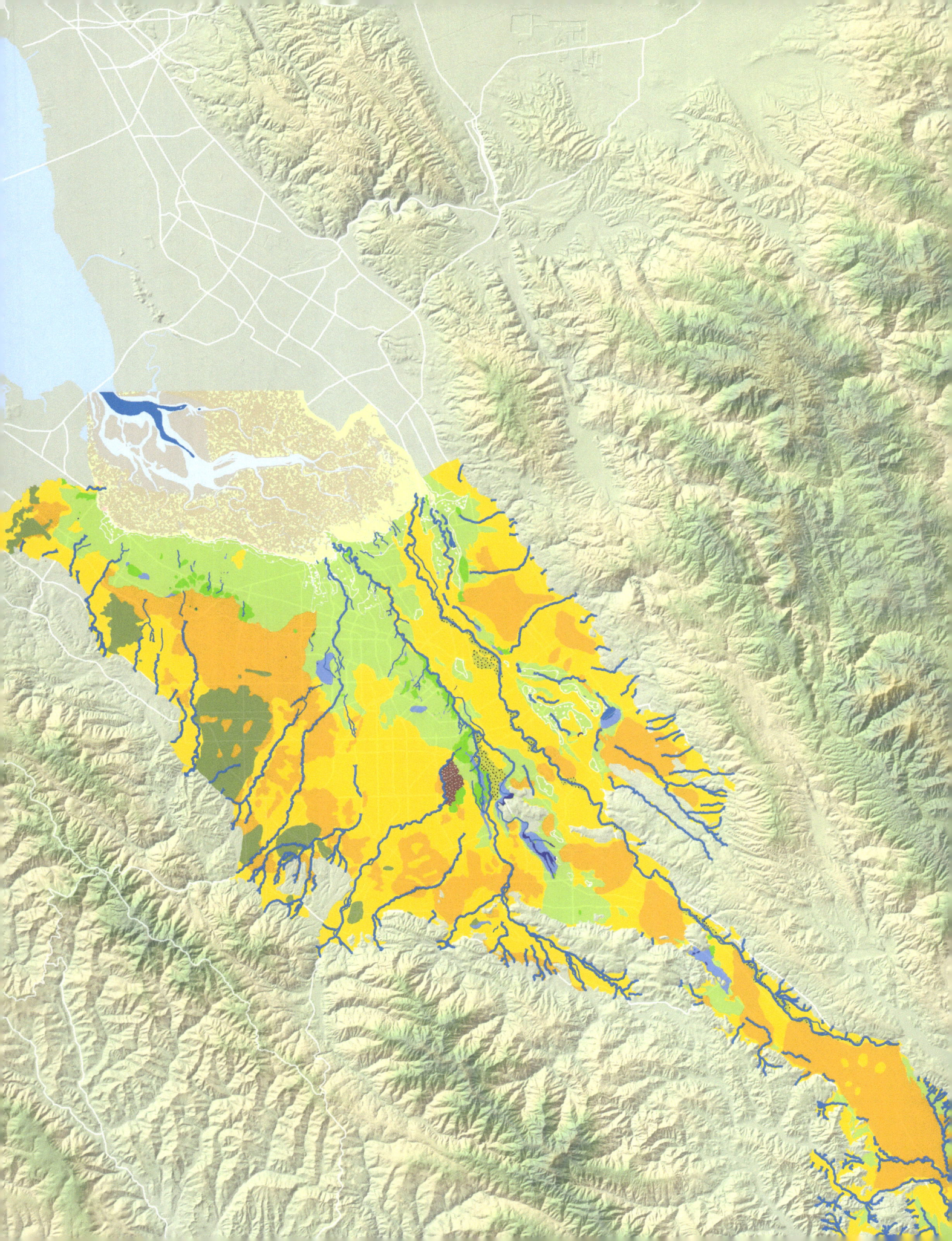

Over the intervening centuries, the Santa Clara Valley has transformed again and yet again: from *Llano de los Robles* to the highly productive agricultural region known as the "Valley of Heart's Delight" to the internationally renowned Silicon Valley. Over this time, former oak woodlands have become subdivisions and seasonally flooded meadows have been converted to office parks. Tidal marshes have been leveed and diked to create salt ponds and housing tracts, while creeks have been ditched and straightened. Once-coherent habitat mosaics have experienced loss and fragmentation; only traces of the historical landscape remain.

Despite these dramatic modifications, the Silicon Valley landscape has retained significant habitat for plants and animals. Large areas of protected open space, ranches and farmland, urban parks, and natural areas all continue to support diverse suites of native species and communities, including unique endemic species and communities such as the Ridgway's rail, Bay checkerspot butterfly, California red-legged frog, western pond turtle, western burrowing owl, sycamore alluvial woodland, valley oak woodland, mixed serpentine chaparral, and many others. In total, Santa Clara County supports hundreds of species of mammals, birds, reptiles, amphibians, and freshwater fish, as well as considerable native invertebrate and plant diversity (ICF International 2012). A number of ongoing efforts are working to protect, enhance, and restore Silicon Valley ecosystems. These include the Santa Clara Valley Open Space Authority's open space conservation efforts, the South Bay Salt Pond Restoration Project, and the Santa Clara Valley Water District and other entities' stream restoration activities, along with many others.

SAN
FRANCISCO
BAY

1:250,000

5 miles

2014 NAIP imagery, courtesy USDA

ELEMENTS
of a Resilient Silicon Valley Vision

The following pages outline a vision for increasing landscape resilience for each of Silicon Valley's four major landscape units: creeks, hills, valley floor/urban matrix, and baylands. Short descriptions of each region, along with lists of key ecological functions of interest and primary stressors of concern, are provided for context.

Several recurring themes emerge from the elements compiled for each of these areas. Large areas of protected open space and habitat are important across the entire landscape: they would provide sufficient space and resources to support large and genetically diverse populations, allow species and habitats to shift as conditions change, and accommodate large-scale landscape processes. At the landscape scale, diversity in the type, distribution, and spatial configuration of these habitats is critical to provide a range of options for native species. Small-scale variations and heterogeneity within each habitat – for example, in topography, salinity, groundwater levels, or vegetation height – would provide areas protected from heat, flood, drought, or high tides that would serve as refuges for wildlife. Linkages between habitats would allow species to move, providing access to resources and options for places to go as conditions change. Natural or naturalistic physical processes and disturbances, such as fire and grazing in the hills, flooding and sediment delivery along creeks, and tidal action in the baylands, would create and sustain this habitat diversity and heterogeneity over time. Collaborative and adaptive landscape management is crucial for capitalizing on emerging opportunities, incorporating lessons from previous projects and advancing science, and preventing short-sighted management decisions that could reduce landscape resilience in the long term.

Cumulatively, these landscape elements would create a network of large, connected, and diverse habitats that are sustained by key processes. A diverse mosaic of forest, grassland, scrub and chaparral habitats would blanket the hills, spanning elevation and temperature gradients and providing habitat and movement corridors for large mammals such as coyotes, bobcats, and mountain lions between the Santa Cruz and Diablo ranges. In the baylands, tidal marsh, mudflats, salt ponds, and other habitats would support healthy populations of shorebirds, waterfowl, salt marsh harvest mice, and other wildlife, while broad transition zones and topographic heterogeneity within the marsh would provide refuge for wildlife from flooding and support the persistence of tidal marsh as sea levels rise. Between the hills and the baylands, the urbanized valley floor would support a broad array of native plants and wildlife through native landscaping and green infrastructure: features such as groves of oaks in parks, backyards, and campuses; wildflowers along street medians; and freshwater wetlands and willow groves alongside stormwater retention basins would provide pathways across the developed landscape for mammals and migratory birds as well as habitat for species such as pollinators, lizards, songbirds and other small animals within the urban matrix. Streams would connect across all of these habitats from the headwaters to the bay through continuous ribbons of streamside vegetation, supporting broad areas of habitat for riparian birds, steelhead and resident fish, and protected movement corridors for wildlife.

In addition to benefits to wildlife, such a landscape would transform our own experience by integrating nature into the developed landscape and increasing our access to immersive, beautiful, and wildlife-filled natural areas. It would also provide a number of benefits and services to society: for example, tidal marshes and transition zones would buffer against rising sea levels; floodplain habitats would provide flood protection for cities, contribute to water supply, and improve water quality through groundwater infiltration and recharge; and oaks would modulate temperatures and provide shade in cities.

More detailed descriptions of these landscape elements are listed in the following section. Note that we frequently suggest "appropriate" or "sufficient" quantities of a particular landscape feature, or refer to "important" or "key" species, communities, or processes. These generic terms are used as placeholders; they will be specified and quantified during the next phase of the project by drawing on local scientific data and in consultation with local experts and advisors.

REGIONAL

Silicon Valley is home to millions of people, along with a wide diversity of habitats and species. It spans a broad gradient from developed to wild landscapes. The vision elements listed below pertain to the Santa Clara County watersheds that drain to the San Francisco Bay, from the hills down to the baylands. Here we identify considerations applicable at the broadest regional scale.

An ecologically resilient Silicon Valley includes...

- Planning and management targets for ecological function and biodiversity informed by a scientific understanding of **what makes Silicon Valley unique:** its past, present, and potential biological communities, geophysical context and drivers, and land use trajectories

- **Large areas of protected open space and habitat** for native species, primarily in the hills and the baylands

- **Diverse habitat mosaics** that contribute to biodiversity and ecological function at a variety of scales

  » Support for a high diversity of species, including unique or rare species and communities that contribute to **regional and global biodiversity** (e.g., serpentine communities, sycamore-alluvial woodland, valley oak woodland, salt marsh endemic birds and mammals)

  » Habitats expressed across important physical **gradients** (e.g., in moisture, elevation, and salinity)

  » **Within-habitat** structural complexity and physical heterogeneity (e.g., in vegetative structure, topography, groundwater levels, and soils), to provide wildlife refuges, promote and sustain genetic and phenotypic diversity and alternative life history strategies, and create high-tide, thermal, flood, and drought refuges

- **Sufficient water availability** in a semi-arid climate to maintain streams, wetlands, and areas with near-surface groundwater and provide water for the plants and animals using these habitats

- Unimpeded or naturalistic **physical processes** and conditions needed to create and sustain habitat heterogeneity (e.g., fires and grazing that sustain grassland habitats, floods that connect channels to floodplains and deliver sediment to marshes)

- **Functional connectivity** across the landscape

    » **Connections** between terrestrial habitats in the hills, tidal habitats in the baylands, and between the hills and baylands (via permeability across the developed urban matrix and through stream riparian corridors), where appropriate, to allow for gene flow, wildlife movement and migration, shifts in habitat location, and transport of water and sediment over time

    » Habitats **isolated** where appropriate (i.e., not over-connected) to create discrete patches that support distinct populations, promote redundancy, and reduce susceptibility to stressors (e.g., disease, invasive species, and catastrophic disturbance from fire or flood)

- **Built environment** that contributes to regional biodiversity and ecological function through habitat restoration, widespread planting of native and locally appropriate vegetation, green infrastructure, and low-impact development

- **Public engagement and investment in ecological resilience and adaptive management,** planning, and design that considers long time scales (see "How do we get there?", page 32)

STREAMS
and riparian habitat

Stream networks within Silicon Valley drain watersheds of varying sizes and support a diversity of creek and riparian habitats, including perennial and intermittent reaches, riparian forest, riparian scrub, deep pools and floodplains.

KEY ECOLOGICAL FUNCTIONS INCLUDE:

- providing habitat and resources for resident creek fish (e.g., California roach, hitch, prickly sculpin, riffle sculpin), aquatic invertebrates, riparian birds (including neo-tropical migrants) and amphibians (e.g., Pacific chorus frogs, California red-legged frog)

- migration and spawning habitat for anadromous fish (e.g., steelhead)

- movement corridors for mammals (e.g., mule deer, coyote, and bobcat)

- stream shading

- improving water quality, biogeochemical cycling of nutrients

- phytoplankton and macroinvertebrate productivity

- sediment transport and storage

- shallow subsurface groundwater storage

PRIMARY STRESSORS INCLUDE:

- increased temperature

- increased drought frequency

- increased fire intensity

- changes in precipitation

- extreme flooding

- sea level rise (at stream mouth)

- impacts from urban development (e.g., habitat loss and fragmentation, pollution), lack of coarse sediment supply, groundwater extraction, urban runoff, increased peak flows, invasive species

above left, imagery courtesy Google Earth

An ecologically resilient Silicon Valley includes...

- **Stream flows with naturalistic magnitude, timing, and duration** to support habitat diversity, transport sediment, and maintain natural cues for fish and other aquatic and riparian organisms

  » **Heterogeneity in surface flow,** including perennial, intermittent, and ephemeral reaches restored and maintained where appropriate to support a range of species and as a barrier to the spread of invasive species (such as bullfrogs and non-native fish), coupled with wet-season connectivity so fish (particularly steelhead) can get to upper reaches and spring-season connectivity for out-migrating steelhead

  » **Floods** managed to support riparian habitat complexity and diversity, promote groundwater recharge, and deliver sediment and woody debris to creeks and baylands

  » Flows that cue **germination** of sycamores, willows, and other native riparian species in appropriate locations; flows that cue steelhead and other **fish up-migration, spawning, rearing and out-migration**

- Increased **sediment transport and delivery** from upper watersheds to channel, floodplain, and baylands

  » Sufficient **coarse sediment** (gravel, cobbles and boulders) to creeks to sustain aquatic habitat (e.g., to support steelhead populations) and avoid accumulation of excessive fine sediment

  » **Sufficient fine sediment** to baylands to support tidal marsh persistence

- **Floodplains of sufficient width** and connection to channel to promote groundwater recharge; support riparian habitat; provide habitat and food for wildlife; and accommodate extreme flooding, rising sea levels, and geomorphic dynamism (including geomorphic responses to climate change and urbanization)

- A diversity of **riparian and floodplain habitat types,** including regionally rare types (e.g., sycamore-alluvial woodland), that provide habitat and help recharge groundwater

  » Habitats sustained through **scour and deposition** wherever possible

  » Connected via **continuous riparian corridors** where appropriate for terrestrial and riparian wildlife movement, in places connecting the hills to the bay

  » **Connectivity between foothill and alluvial stream reaches (lack of barriers)** for steelhead and resident fish, including access to suitable habitat upstream of dams where possible

- Microclimates, microtopography, complex vegetative structure, coarse woody debris in channels, and other physical **heterogeneity** within riparian habitats to support refuges

  » **Thermal and drought refuges** to mitigate hotter water temperatures and drier summers (e.g., shaded riparian cover, hyporheic flow, deep stream pools, cold-water inputs)

  » **Flood refuges** for aquatic organisms, especially fishes, to escape high velocity flows in urban stream channels and for terrestrial species (e.g., voles) to escape flood waters

- **Sustainable management** of water, sediment, and land use to achieve heterogeneity and refuge habitat

  » Management of **stormwater flows** to promote groundwater recharge

  » **Reservoir** operation (e.g., reservoir redesign or changes in flow releases) or dam removal to support sediment transport and flood pulses, as well as reliable perennial reaches

  » Levee setbacks, daylighting of creeks, retreat or removal of development, and/or floodplain grading to support **floodplain habitat** restoration and hydrologic reconnection to channel

  » **Groundwater levels maintained** through recharge due to improved grazing practices, low-impact development (LID) to reduce permeable surfaces, and regulation of groundwater extraction

STREAMS

HILLS

Mountain ranges and foothills flank Silicon Valley to the east (Diablo Range) and west (Santa Cruz Range), spanning significant gradients in elevation and latitude. Current land uses include parks and open space supporting diverse habitat types such grassland, oak woodland, forest, and scrubland/chaparral, along with limited residential development and agricultural and ranch land.

KEY ECOLOGICAL FUNCTIONS INCLUDE:

- providing habitat and resources for terrestrial vertebrates (e.g., coyote, bobcat, southern alligator lizards, common kingsnakes, acorn woodpeckers) and native plants (e.g., blue oaks, valley oaks, native grass and chaparral communities)

- movement of terrestrial wildlife species across large open spaces

- natural watershed processes in headwater areas (e.g., reduced erosion, natural percolation/infiltration of rainwater, recharge of local water supplies)

PRIMARY STRESSORS INCLUDE:

- increased temperature

- decreased water availability (e.g., groundwater, soil moisture, and precipitation)

- increased drought frequency

- increased wildfire frequency

- decreased fire frequency

- nitrogen deposition

- invasive species

- encroachment of urban development

above left, imagery courtesy Google Earth

An ecologically resilient Silicon Valley includes...

- Large areas of **protected open space,** particularly in areas identified as high conservation priorities

- **Functional connectivity** among upland habitats and open space areas to support gene flow and dispersal of individuals

    » Connectivity **between the Diablo and Santa Cruz ranges** at the southern edge of Silicon Valley for wildlife movement (e.g., movement of large mammals including coyote, deer, bobcats, and mountain lions)

    » Connectivity **up and down mountain ranges** to provide room for plants and habitats to shift in response to climate change

- A **diversity of habitats,** vegetative communities, and conditions, including chaparral, grassland, forest, and scrub, spanning a range of climate conditions

    » **Rare, vulnerable habitat types,** including redwood forests (threatened by changing climate conditions) and serpentine grasslands (threatened by shrubland encroachment and invasive plants)

    » Hot and/or dry areas occupied by **drought-tolerant native vegetation** (e.g., blue oaks, chamise, knobcone pines) that could serve as seed sources for the future, including dry microclimates (for dispersal) and areas at the southernmost end of the valley

» Wildlife support in **agricultural areas and ranchland** (e.g., food, cover, perches for birds)

» Diverse **genotypes** available to maximize the chances of successful response to stressors

- Relatively **cool areas maintained by microtopography** and/or microclimates that provide temperature refuges via proximity to water, shading, topography, hillslope/aspect, fog, and coastal influence

- Areas with relatively **high groundwater and/or reliable soil moisture** maintained by aquifer recharge to provide drought refuges

- Management of **upper portions of watersheds** to maintain water quality, adequate flows, and sediment supply downstream

- Appropriate levels of disturbances such as **fire and grazing,** or human management such as thinning or clearing that mimics these processes, to support a heterogeneous matrix of grassland, shrubland, woodland, and chaparral

- Subdivision **zoning and fire breaks/buffers** that allow for large wildland areas to burn with reduced risk to the built environment

- **Post-disaster management plans** (e.g., to guide re-planting, re-seeding, stressor management or other actions after fires) plus availability of **nursery stock** of preferred species

HILLS

VALLEY FLOOR

The Silicon Valley floor is characterized by a gently to moderately sloping alluvial fan that historically supported a variety of wetland and terrestrial habitats, including wet meadows, grasslands, and oak woodlands. Current land uses are predominantly commercial and residential development.

KEY ECOLOGICAL FUNCTIONS INCLUDE:

- providing habitat and resources for native invertebrates, particularly pollinators and soil aerators, amphibians (e.g., Pacific chorus frogs), reptiles (e.g., Western fence lizard), mid-size predators (e.g., grey fox, coyote, bobcat), and migratory or highly mobile species (e.g., songbirds, raptors, waterfowl, bats)

- support for native plant diversity via urban landscaping

- support for willow groves and other groundwater supported wetlands

- groundwater recharge and storage

PRIMARY STRESSORS INCLUDE:

- urban development (habitat loss and degradation, barriers to connectivity, non-native and invasive plants, pollution)

- decreased water availability

- increased temperature

- increased drought frequency

- non-native and nuisance predators (e.g., feral cats)

- human disturbance

- polluted runoff

above left, imagery courtesy Google Earth

An ecologically resilient Silicon Valley includes...

- **Space for key physical processes,** notably flooding and sediment transport, to be accommodated by urban stream channels wherever possible (e.g., through undeveloped floodplains, levee setbacks)

- Streams with **functionally connected patches of riparian habitat** that serve as wildlife movement corridors between the hills and the bay (e.g., bats, songbirds, and grey foxes)

- A diversity of **wetland habitats** (e.g., perennial, intermittent, and ephemeral streams; willow groves; depressional wetlands; seasonal grasslands; treatment ponds; and rain gardens) to support plants and wildlife in a water-limited environment

- **Native landscaping** with high species diversity, structural complexity, and sufficient patch size to provide habitat and connectivity for native species

  » Selection of **local species and genotypes,** with consideration for including other genotypes (or selection from within the existing range of variation), based on climate change projections

  » Native planting palettes that include **species likely to tolerate heat and drought stresses** (e.g., valley oak, toyon, and coffeeberry)

  » **Application** in parks, backyards, greenways, medians, sidewalks, office parks, and other locations to provide habitat and permeability across the developed landscape

» **Complex vegetative structure,** including high density, overlap, height diversity, and structural variation

» **Coordinated planting efforts** that add up to habitat at a landscape scale (e.g., "re-oaking" on individual sites to mimic densities of historic oak woodland/savanna)

- **Green infrastructure and low-impact development (LID)** that support or mimic natural physical processes and/or provide habitat (e.g., rain gardens and retention basins to support recharge for groundwater-dependent habitats)

- **Removal of barriers to wildlife movement and reduction of sources of mortality** where feasible, taking advantage of opportunities based on infrastructure and landscaping updates (e.g., road underpasses and overpasses for wildlife, removal of fencing, reduced contaminants in runoff, plantings with predator-exclusion zones such as bramble thickets, integrated pest management)

- Wetland complexes in areas supported by appropriate **soils and topography,** with **groundwater levels** sufficient to maintain groundwater-dependent habitats (e.g., including willow groves; persistent, stratified summer stream **pools** for aquatic organisms; and naturally **perennial stream reaches)**

- **Buffers** between wildlands and developed areas to protect from human encroachment and non-native and nuisance predators

- **Predation pressure from non-native and nuisance species controlled** in ways that limit harm to native species, with limited and careful use of herbicides and pesticides

VALLEY

BAYLANDS
and estuarine-terrestrial transition zone

Tidally influenced bayland habitats ring the bay, supporting extensive tidal marsh, salt pond, mud flat, and other intertidal habitats, including large restoration projects. The estuarine-terrestrial transition zone that connects the baylands to adjacent upland habitats was historically varied but today consists mostly of steep levees.

KEY ECOLOGICAL FUNCTIONS INCLUDE:

- providing habitat and resources for rare and endemic marsh species including Ridgway's rail and salt marsh harvest mouse

- nursery and foraging habitat for estuarine and anadromous fish

- overwintering, migratory stopover, and breeding habitat for waterbirds

- flood protection and erosion prevention

- primary productivity

PRIMARY STRESSORS INCLUDE:

- sea level rise

- development

- sediment limitation

- invasive species

above left, imagery courtesy Google Earth

An ecologically resilient Silicon Valley includes...

- Large areas of protected bayland habitat (e.g., National Wildlife Refuge and South Bay Salt Pond Restoration Project)

- Diverse bayland habitat mosaics that include a variety of subtidal and intertidal habitat types (e.g., subtidal channels, tidal marsh, mudflats, oyster and eelgrass beds, salt ponds, complex channel networks, and transition zone), as appropriate for the local setting

- Connectivity between bayland habitats and to appropriate surrounding uplands for wildlife movement around the perimeter of the Bay, through tidal marsh and transition zone patches and corridors

- Sufficient sediment delivered to the baylands from local watersheds or other sources; "elevation capital" preserved where possible to support long-term tidal marsh persistence

- Channel complexity and topographic heterogeneity within baylands to provide habitat for diverse species

  » Critical low marsh vegetation, including native cordgrass; submerged aquatic vegetation including eelgrass beds, sago pondweed and widgeon grass

» Topographic highs within tidal habitats and gradual transitional zones between tidal and terrestrial habitats, where appropriate, to provide high-tide refuges (e.g., for salt marsh harvest mice and Ridgway's rails)

• Infrastructure and assets moved out of flood hazard zones allowing accommodation space for landward marsh migration as sea levels rise

• Habitats expressed across appropriate physical gradients (e.g., salinity and elevation)

BAYLANDS

Stevens Creek Trail, Mountain View (photo courtesy Stanislav Sedov February 2014)

How do we get there?

This vision for landscape resilience can help highlight conservation, restoration, design, and management priorities. No matter how clearly articulated the vision, however, it must be accompanied by a strategy for how to achieve the vision through collaborative and adaptive landscape management. The vision must be flexible enough to allow implementers to take advantage of new opportunities and incorporate new ideas as they arise. It must also be shared and implemented by a broad range of stakeholders, including scientists, managers, planners, and citizens. Multiple, coordinated actions involving a variety of landowners and stakeholders must be executed at various spatial and temporal scales, from urban tree planting and flood control channel redesign to open space conservation and rangeland management. While this will not be easy, it will yield an enormous reward – of conservation, restoration, urban design, and management dollars cumulatively contributing to more resilient Silicon Valley ecosystems.

The following bullets represent some of the most important considerations in managing for landscape resilience. Additional suggestions for effective stewardship, education, and adaptive management strategies are detailed in other regional planning and goals documents (e.g., Santa Clara Valley Greenprint, Conservation Lands Network).

Managing for a resilient Silicon Valley involves...

- Planning with **long time-scales** (multi-decadal) in mind; **short-term actions** that support (and don't preclude) long-range resilience planning (e.g., zoning, sediment management, and restoration design) rather than exacerbating stressors, depleting critical resources, or fundamentally altering critical physical processes or settings

- **Sustained and adequate funding** for conservation, restoration, planning, and management actions

- **Research, monitoring, and pilot projects** incorporated into planning, management and restoration activities (e.g., monitoring of habitat extent and composition, wildlife indicator populations, pilot projects on reservoir or stream hydrograph management)

- Emergency and **disaster response plans** and preparedness to take advantage of potential opportunities after extreme events (e.g., post-flood riparian restoration plans that take advantage of natural regeneration processes; post-fire management plans that anticipate opportunities for revegetation with appropriate nursery stock), rather than returning to business as usual

- **Contingency funds** set aside for rapid post-disaster response

- "Mainstreaming" the resilience vision into **capital improvement plans** to take advantage of the redesign of first-generation infrastructure (e.g., flood-control channel redesign, road improvements) and subsequent regeneration cycles

- Active management of potential **stressors** of greatest concern (e.g., nitrogen deposition in the hills, invasive species such as *Spartina alterniflora);* anticipation of acceleration in some stressors (e.g., sea-level rise) over the next century and beyond

- Active management of **critical resources** (e.g., sediment moving to the Baylands; groundwater levels supporting wetland habitats; topography and water needed to support high-tide, temperature, flood, and drought refuges)

- Ecological **resilience planning** institutionalized through dedicated implementation staff and incorporation into planning documents (e.g., General Plans)

- **Education and outreach** to inform land managers and the public of actions that benefit ecological resilience (e.g., xeric landscaping, control of cats, limiting pesticide use)

- **Coordination and partnerships** among planning efforts, agencies, and other stakeholders

- Opportunities for people to interact with nature in a way that educates and inspires **financial and emotional investment** in ecosystems and good stewardship

References

Bay Area Open Space Council. 2011. *The Conservation Lands Network: San Francisco Bay Area Upland Habitat Goals Project Report.* Berkeley, CA.

Beller E, Robinson A, Grossinger R, Grenier L. 2015. *Landscape resilience framework: operationalizing ecological resilience at the landscape scale.* SFEI contribution #752. San Francisco Estuary Institute, Richmond, CA. Available at resilientsv.sfei.org.

Beller E, Salomon M, Grossinger R. 2010. *Historical vegetation and drainage patterns of western Santa Clara Valley: a technical memorandum describing landscape ecology in Lower Peninsula, West Valley, and Guadalupe Watershed Management Areas.* SFEI contribution #622. San Francisco Estuary Institute, Oakland, CA.

Committee for Green Foothills. 2014. *Deep roots, green future: an environmental vision for the next 50 years.* Available at www.greenfoothills.org/vision/.

County of Santa Clara Office of Sustainability. 2015. *Silicon Valley 2.0 climate adaptation guidebook.* See www.sccgov.org/sites/osp/SV2/Pages/SV2.aspx.

Goals Project. 2015. The Baylands and Climate Change: What We Can Do. The 2015 Science Update to the Baylands Ecosystem Habitat Goals prepared by the San Francisco Bay Area Wetlands Ecosystem Goals Project. California State Coastal Conservancy, Oakland, CA.

Grossinger R, Askevold R, Striplen C, et al. 2006. *Coyote Creek Watershed Historical Ecology Study: historical condition, landscape change, and restoration potential in the Eastern Santa Clara Valley, California.* SFEI contribution #426, San Francisco Estuary Institute, Oakland, CA.

Grossinger R, Striplen C, Askevold R, et al. 2007. Historical landscape ecology of an urbanized California valley: wetlands and woodlands in the Santa Clara Valley. *Landscape Ecology* 22:103-120.

ICF International. 2012. *Final Santa Clara Valley Habitat Plan.* San Francisco, CA.

Santa Clara Valley Open Space Authority. 2014. *The Santa Clara Valley Greenprint: A guide for protecting open space and livable communities.* San Jose, CA.

Acknowledgments

This document was developed as part of Resilient Silicon Valley, a project funded by a charitable contribution from Google's Ecology Program. For more information on Resilient Silicon Valley, please visit resilientsv.sfei.org.

Thanks to our Silicon Valley Regional Science Advisory Team for their involvement in the project, including contributions to and review of this document. We would also like to give thanks to the Google Book Club, a working group of local environmental NGOs and Google employees, who helped inspire and catalyze this work. We would like to give special thanks to Audrey Davenport, Google's Ecology Program lead, whose leadership and vision have made this effort possible.

Appendix:
Resilience Worksheets

The worksheets reproduced below were used to guide us through an initial application of our Landscape Resilience Framework to Silicon Valley. They represent responses to the question, "What are the elements of an ecologically resilient Silicon Valley landscape?" The most important elements were then synthesized in collaboration with our Regional Science Advisory Team into the draft vision described above.

The landscape elements included in the worksheets represent a working draft of potential strategies to consider as derived from the resilience framework, not a finalized list of vision recommendations. We include them here to give a sense of the brainstorming process and breadth of possible elements that were considered. The worksheets have been left in draft form and have not been edited for content.

SETTING

Component	Definition	An ecologically resilient Silicon Valley landscape includes....
Geophysical context	Underlying geology, soils, hydrology, and topography key to the feature or site's identity and persistence	• Preserved intact or restored **soils, topography, and groundwater** support appropriate habitats (e.g., depressional wetlands in low areas with high groundwater levels and clay soils, serpentine soils for serpentine grasslands, coarse alluvial soils for oaks; groundwater levels sufficient to maintain persistent, stratified summer stream pools for aquatic organisms and naturally perennial stream reaches)
Ecological context	Ecological assemblies; dominant and rare/unique vegetative communities that distinctively characterize the landscape. Includes landscape legacies – remnants of former populations, habitats, structures, and processes that can be preserved, built on, or learned from/used as analogs	• **Dominant native vegetative communities** present at sufficient scale to persist • Locally **rare key native vegetative communities** present and currently extirpated communities restored at sufficient scale to persist (e.g., oak savanna and woodlands, serpentine grasslands, sycamore-alluvial woodland) • **Remnant** habitats and habitat elements integrated into landscape (e.g., heritage oaks and other older/larger trees, standing dead and fallen trees, alkali grasslands, remnant tidal marshes, tidal channels in salt ponds, rock outcrops and cliffs) • Restoration and management of areas where **native ecological communities** can feasibly be restored (e.g. tidal marsh, willow groves, serpentine grasslands) • **Hybrid and novel** systems support ecological functions where historical habitats are not recoverable (e.g. annual grasslands, estuarine benthos) • Natural **ecosystem processes sufficiently intact** to support self-sustaining natural habitats and communities in key areas
Historical/ cultural context	How the landscape has changed over time – which ecosystem elements have persisted or disappeared, and why	• Restoration and management based on an understanding of **local history** and change over time (e.g., composition and width of former t-zone habitat informs t-zone restoration, understanding of historical composition and distribution of oak woodlands guides re-oaking) • Restoration and management based on an understanding of **potential future trajectories** and opportunities as infrastructure and landscapes are redesigned • Restoration and management based on **local knowledge** of how to sustainably steward landscapes and ecosystems (e.g. TEK on fire and oak management)
Critical resources	Resources required for the persistence of desired ecological functions but currently limited within the landscape	• Variable **aquatic and wetland habitats** (e.g. perennial, intermittent, and ephemeral streams, depressional wetlands, willow groves, high groundwater, treatment ponds, rain gardens) to support plants and wildlife in a water limited environment. • **Thermal and drought refuges** to accommodate hotter temperatures and drier summers (e.g. shaded riparian cover, hyporheic flow, deep stream pools) • **Baylands that receive enough sediment**, via watershed management or other approaches, to support rapid marsh accretion that will offset SLR in a time of declining Bay sediment. • Opportunities for **wildlife support on the valley floor** (where open space/wildlife habitat is limited) in unconventional areas such as landfills, golf courses, institutional lawns, airports, water treatment basins, etc.

PROCESS

Component	Definition	An ecologically resilient Silicon Valley landscape includes....
System drivers	Large-scale forces such as climate change and land use	• Floodplains, flood-prone areas, and shoreline areas below Mean Higher High Water are **not developed** • **Macrotidal aspect** of estuary preserved
Disturbance regimes	Expected but unpredictable events, such as fires, floods, and droughts, that shape habitat structure and/or create opportunities for wildlife	• **Natural and managed disturbances** support habitat complexity and diversity (e.g., fire, manual clearing, and grazing in chaparral/scrub/grassland in hills; floods in channel and adjacent riparian areas) • Natural disturbances are **encouraged by land use and zoning** that keeps development out of fire and flood prone areas. • **Management of successional transitions** that reduce habitat diversity in the absence of disturbance, especially fire (e.g., Douglas fir overtopping oak woodland and coyotebush expanding in grasslands) • **Small-scale, intermediate disturbances** (e.g., gophers) to create microsite heterogeneity
Habitat-sustaining processes	Dynamic processes, such as the transport of water and sediment, that are key to maintaining habitats	• Sufficient fine **sediment delivery** to baylands and floodplains via creek channels, tidal waters, and enhancement projects to sustain marsh and riparian habitats; sufficient coarse sediment delivery to creeks to sustain aquatic habitat • **Naturalistic magnitude and timing of environmental flows** delivered to creeks and across floodplains (e.g., flows that cue germination of sycamores and other native riparian species in appropriate locations, and fish migration, rearing and spawning; avoidance of hydromodification [excessive stormwater flows causing creek erosion]) • Shallow **groundwater levels** where appropriate to support groundwater-dependent habitats (e.g., willow groves, freshwater wetlands, permanent stream pools, naturally perennial stream reaches) • Native and/or managed **grazing** to control excessive plant growth (due to increased anthropogenic nitrogen inputs, invasive species, etc), maintain ecological diversity and groundwater recharge in grassland/savanna • **Reservoir** operation (e.g., reservoir redesign or changes flow releases) or dam removal to support sediment transport and flood pulses, as well as reliable perennial reaches • Maintain high **propagule pressure** of desired species

CONNECTIVITY

Component	Definition	An ecologically resilient Silicon Valley landscape includes....
Linked habitat patches	Habitat distribution supports different aspects of species life history, allows for species movement and migration, exchange of resources, and gene flow between habitat patches (functional connectivity)	• Functional connectivity/permeability between **large open space areas** (especially among upland habitats and among bayland habitats) connected through corridors/stepping stones. • **Hills to baylands** connectivity through streams/riparian corridors and more permeable valley floor (e.g., via urban greening) for wildlife movement, dispersal, and transport of sediment and other materials • **Wetland complexes** that provide a critical stopover (in an area otherwise lacking appropriate habitat) for migratory songbirds along the Pacific Flyway • **Linked channels or floodplains** at the mouths of certain streams • **Removal of barriers** to wildlife movement (e.g., road underpasses and overpasses for wildlife, removal of fencing) • **Connectivity to habitat outside** of the Silicon Valley (e.g., pacific flyway)
Space for species and habitat ranges to shift	Space for species and habitats to move to as their ranges shift, including accommodation space	• **Accommodation space** and transition zone habitats upslope of baylands to allow for landward marsh migration as sea levels rise • **Creek corridors and floodplains** of sufficient width to accommodate current and predicted future flood events and rising sea levels (e.g. via levee setbacks, retreat or removal of development, and/or floodplain grading) • Open space areas and habitat patches of any size throughout the landscape that could serve as **stepping stones** and **seed sources for colonization.** • Hot and/or dry areas occupied by drought-tolerant native vegetation that could serve as seed sources for future • Contiguous open spaces that **cross gradients of varying steepness** (e.g., elevation gradient for upland habitats, salinity gradient for marshes)
Gradual transitions	Soft edges between habitat types that support ecotones	• Gradual **transition zones** between baylands and terrestrial habitats • A diversity of steepness in habitat transitions that includes areas of **non-abrupt transitions/continuum** between grassland and woodland/forest habitats, riparian and upland habitats • **Buffers around wetland and aquatic habitats** to support ecotones (and provide accommodation space and ameliorate stressors)
Expression of habitats across gradients	Expression of habitats across important physical gradients, such as salinity and temperature	• Habitats that span key spatial and physical **gradients in salinity, temperature, and elevation** (e.g., tidal marsh expressed along a salinity gradient, chaparral expressed across a temperature gradient, streams expressed as longitudinal gradient) • Estuarine-terrestrial **transition zone that rings the South Bay** • **Upland habitats across different mountain ranges** (e.g., between Santa Cruz Mountains and Diablo Range) • Expression of upland habitats across the valley with a gradient of **distance from the Bay/coast** • Stream habitats supported across **north/south and east/west gradients** that account for changes in precipitation and temperature.
Landscape coherence	Habitats are organized in a way that supports desired processes and ecosystem functions, including the ability of wildlife to navigate within the landscape	• **Complete ecosystems,** with key components and processes intact at the appropriate scale (e.g., connected bayland habitats, including mudflat, marsh, and T-zone, that follow important physical gradients, align with natural processes, and allow system components to interact in ways that better support wildlife) • **Hydrology** (flow timing, duration, distribution, magnitude, connectivity, etc.) and water chemistry that maintain natural cues for fish and other aquatic and riparian organisms

DIVERSITY/COMPLEXITY

Component	Definition	An ecologically resilient Silicon Valley landscape includes....
Richness of land-scape features	Landscape-scale diversity in habitat types and connections between different habitat types; physical heterogeneity in topography, groundwater, soils	• A **diversity of habitats,** primarily for key native species (e.g., oak savanna/woodland; redwood and mixed forest; serpentine, perennial, and annual grasslands; chaparral; alkali meadow/grassland; willow groves; perennial freshwater wetlands and ponds; seasonal wetlands; riparian forest; sycamore-alluvial woodland; salt marsh, brackish marsh, salt flats, mud flats, T-zone, and ephemeral, intermittent and perennial streams) • Variable, **heterogeneous topography** to support habitats and species of interest (e.g., intact low topography to support depressional wetlands) • **Variable aquatic and wetland habitats** (e.g. perennial, intermittent and ephemeral streams, depressional wetlands, willow groves, high groundwater, treatment ponds, rain gardens) to support plants and wildlife in a water limited environment • Management of **successional transitions** that reduce habitat diversity in the absence of disturbance, especially fire (e.g., doug-fir overtopping oak woodland and coyotebush expanding in grasslands)
Within-habitat diversity and complexity	Site- or habitat-scale vegetative diversity (e.g., in species, structures, or height) and physical heterogeneity (e.g., in microhabitats, microtopography, and microclimates)	• Microclimates, microtopography, complex vegetative structure, and other **heterogeneity** supporting within-habitat complexity to provide wildlife refuge and promote/sustain genetic and phenotypic diversity and alternative life history strategies (e.g., stream pools, pannes, channels of varying size in marsh; complex understory creating light and temperature gradients in scrub/riparian) • Within-habitat **diversity in vegetation** age structure, vertical structure, composition and spatial configuration (e.g. oaks of varying ages, different densities in chaparral) • Areas that provide **temperature refuges** in upland habitats (e.g. because of proximity to water, shading, hillslope/aspect, coastal influence, etc.) • **High tide refuges** with sufficient space for bayland species to escape flooding conditions • **Riverine flood refuges** to provide aquatic organisms, especially fishes in urban stream channels, with shelter from extreme floods (e.g. wide floodplains, small tributaries) • **Temporal variability** in resource availability (e.g., Plants with a diversity of flowering timing to support a diverse suite in pollinators as life history timing changes; wetland areas that pond at different times of year, intermittent and ephemeral streams) • **Native species assemblages conserved** in invaded annual grasslands
Diversity in approach	Maintaining response diversity and a diversity of life history strategies both within and between species to deal with variability, disturbance, stressors	• Presence of **population segments that use the landscape in different ways** (e.g. rainbow trout/steelhead) • Species that **respond to similar stressors in different ways** (e.g. fire re-sprouters vs. fire-germinating seeds)
Genetic and phenotypic variability	Diversity in genes and traits within species	• Sufficiently **large populations** of key species to support genetic and phenotypic diversity (e.g., Bay checkerspot butterfly, steelhead and rainbow trout) • **Landscape/streamscape complexity** to produce areas that support different species and populations • Endemic **rare and endangered species** (e.g., Bay checkerspot, tiger salamander, red/foothill yellow-legged frog, burrowing owl, least bell's vireo, tri-colored blackbird). • Diverse **seed banks** (and range of genotypes) to support native plant persistence

REDUNDANCY

Component	Definition	An ecologically resilient Silicon Valley landscape includes....
Structural/ spatial	Multiple habitat patches and an abundance of key structures within habitats	• Multiple **habitat patches** providing similar or overlapping functions (e.g., multiple willow grove-wetland complexes, multiple variable depth pools in creeks)) • Multiple **corridors** for wildlife movement (e.g., multiple continuous riparian and non-riparian corridors) • Flows **supporting steelhead runs on multiple streams**, ideally originating from watersheds experiencing different physical gradients, to minimize risk (e.g., Diablo Range vs. Santa Cruz Mountains)
Population	Distinct or disconnected populations of a species	• Enough **distance between population segments** of important species to diversify risk (e.g., high Ridgway's rail densities in more than one marsh patch) • **Multiple streams support freshwater fish populations**, reducing the likelihood of regional extirpation
Functional	Multiple species within the system supporting the same ecological function	• Support for multiple species supporting similar **key functions** (e.g., pollinators, burrowing mammals in grasslands)
Discreteness	Isolation or disruption between habitat elements to reduce susceptibility to stressors	• Risk diversified by **disconnection** (e.g., perennial stream reaches seasonally separated by intermittent reaches) • **Reduction in inter-basin water transfers** • **Breaks in habitat continuity** to provide fire breaks and barriers to the spread of other stressors and disturbances (e.g. breaks in chaparral, not over-connecting marsh channels via ditches)

SCALE

Component	Definition	An ecologically resilient Silicon Valley landscape includes....
Large spaces	Areas of sufficient size to accommodate sustaining physical processes and support sufficiently large wildlife populations and support genotypic and phenotypic variability	• Large areas of **open space** in the hills and baylands to support wildlife (large is defined relative to the species intended to support) • **Floodplains and riparian corridors** of sufficient width to support wildlife and accommodate flooding and geomorphic dynamism (including geomorphic responses to climate change and urbanization) • Large areas of **tidal marsh** and other bayland habitats to support wildlife
Long time scales	Broad time horizons over which ecological functions must persist under changing and variable conditions	• **Accommodation space** and transition zone habitats to anticipate landward migration of tidal marsh as sea levels rise • **Key species and habitats established early** in areas that are likely to support their persistence but not establishment under future conditions (e.g., oaks established while conditions can still support seedlings, marsh restored while sediment supplies are adequate) • **Availability of seed stores and seedlings** for vegetation communities not currently present/abundant but likely to withstand/thrive under future conditions. • Land use and zoning planned with **time horizon for future changes**
Cross-scale interactions	Important interactions that occur across multiple spatial and temporal scales	• Short term and fine-scale actions and visions that **link to long term and large-scale planning** and visions (e.g., preserving remnant habitat near areas likely to be available for restoration in the future) • A balance of resources in the landscape to **account for trade-offs** that happen at different spatial and temporal scales (e.g., a landscape needs large habitat patches, but not at the expense of having no habitat redundancy; habitat diversity but not to the extent that specific critical resources cannot be maintained in adequate abundance)

PEOPLE

Component	Definition	An ecologically resilient Silicon Valley landscape includes....
Ecological engagement	Place-based and widespread landscape stewardship	• Opportunities for people to **interact with nature** in a way that educates and inspires financial and emotional investment in ecosystems and good stewardship • People with a **deep understanding of place** that can inform stewardship strategies • **Coordination and partnerships** among planning efforts, agencies, and other stakeholders
Landscape integration	Opportunities to support ecological functions occur across urban, suburban, agricultural and open space lands	• Habitat integrated into **urban and suburban areas for wildlife support** (e.g., through green buildings, landscaping, parks, backyards, street trees, and riparian corridors). • Habitat integrated into developed areas in a way that leverages large areas and maintains landscape **permeability.** • Rain gardens, retention basins and other **water infrastructure that links to larger regional wildlife support and physical processes** (e.g. leverage recharge in urban and suburban landscape to support groundwater-dependent habitats) • Wildlife support in **agricultural areas and ranchland** (e.g. food, cover, perches for birds provided in hedgerow buffers) • **Landscaping of developed areas** using native vegetation that provides habitat for key species
Adaptive management	Stewardship of the land in a coordinated, flexible, and informed manner; learning from monitoring, research, and pilot projects	• Pilot projects, **novel approaches**, willingness to fail and learn from mistakes • Learning through programs that support **research and monitoring** needed to make informed decisions and actions • **Flexible governance** structure and mechanisms for coordination between stakeholders; incorporation of resilience tenets into **documents** such as general plans • Early detection networks and ways of **rapidly responding to catastrophes or novel stressors** (e.g. new invasives with high potential for harm, advance planning to improve chances of rapid and ecologically effective response to catastrophes; planning for post-fire restoration/management) • Ability to **respond quickly** when unforeseen opportunities arise or catastrophic events occur
Stressor management	Management of specific stressors that must be controlled in order to maintain desired ecological functions and biological processes	• **Buffers** between wildlands and developed areas • Control of **excess nutrients** via source control, consumption of excess production, or other means (e.g. grazing in grasslands, buffers in wetland/aquatic systems, LID) • Feral cat colonies and other **nuisance species** are relocated far from core wildlife habitat • Control of **invasive species** that threaten the persistence of desired ecological functions and desired species • **Reduction in contaminants** that contribute to species mortality (e.g. reduced contaminants in runoff, remediation, green building, etc) • Reduced emissions to **manage nitrogen deposition** or control of the resulting increase in biomass production to preserve serpentine grassland communities • Native and/or managed **grazing** to reduce carbon accumulation where excessive due to nitrogen, grazing, etc.

www.ingramcontent.com/pod-product-compliance
Lightning Source LLC
Chambersburg PA
CBHW041731210326
41598CB00008B/843